Teff in Southern Ethiopia. Effect of Seed Rates and Sowing Methods on Phenology, Growth, Yield and Yield Attributes at Wolaita Sodo

Wolde Tasew

Bibliographic information published by the German National Library:

The German National Library lists this publication in the National Bibliography; detailed bibliographic data are available on the Internet at http://dnb.dnb.de.

ISBN: 9783346421395
This book is also available as an ebook.

Nymphenburger Straße 86
80636 München

Print and binding: Books on Demand GmbH, Norderstedt, Germany
Printed on acid-free paper from responsible sources.

GRIN web shop: https://www.grin.com/document/1023746

Effect of Seed Rates and Sowing Methods on Phenology, Growth, Yield and Yield Attributes of Teff [*Eragrostis Teff* (Zucc.) Trotter] at Wolaita Sodo, Southern Ethiopia

Wolde Tasew,

Wolaita Sodo Agricultural TVET College, Department of Plant Sciences

ABSTRACT:

Teff is a major staple cereal crop in Ethiopia. The need for its production as a staple food is increasing from year to year; however, its productivity is limited, amongst others, by the use of improper sowing methods and inappropriate seed rates. Row planting and optimizing seed rate should be in place to increase the productivity of the crop. A field experiment was conducted at Wolaita Sodo Agricultural College Farm, during the 2018/19 main cropping season under rain-fed condition, with the objective of evaluating the response of teff to seeding rates and sowing methods on phenology, growth, yield components and yield of teff. Factorial combinations of two sowing methods (Row and broadcast) and five seeding rates (2.5, 5, 10, 15 and 20 kg ha^{-1}) were laid out in a randomized complete block design (RCBD) with three replications. Data related to phenology, growth, yield and yield attributes, were collected and analyzed using Gestate software. Plant parameters such as days to panicle emergence, lodging %, number of total and effect tillers per plant and grain yield were significantly affected by the interaction effect of sowing methods and seed rates. Plots sown with low seeding rate (2.5kg ha^{-1}) combined with row planting gave high grain yield (2.5 t ha^{-1}). Therefore, using treatment combinations of seeding rate of 2.5 kg ha^{-1} together with the row planting can be advised for teff production in the study area. However, repeated experiment over different soil type, seasons and locations necessitates for conclusive recommendation.

Key words: Broadcasting, Row planting, Seed Rate, Teff, yield attributes

Contents

1. INTRODUCTION

Teff [*Eragrostis tef* (Zucc.) Trotter] is a cereal crop in belongs to the graminaceae family. It is an indigenous C4, self-pollinated, chasmogamous annual warm season grass that is used throughout Ethiopia as grain crop for human consumption and as forage for livestock [1].It has the largest value in terms of both production and consumption in Ethiopia [2].Teff in Ethiopia stands first in area coverage and second in total annual production next to maize, and ranks the lowest yield compared with other cereals grown in Ethiopia [3]. It is the major staple cereal crops and highly adapted to diverse agro-ecological zones including conditions marginal to the production of most of the other crops[4]. Even though it is small seeded crop, Ethiopian farmers choose to grow teff because of its multiple advantages such as high market value, reduced postharvest management cost, and low risk, and the straw is preferable as animal feed compared with other cereal crops [5].

Despite the preferences and the largest area coverage of teff, its national average yield is very low as compared to other cereals. Teff in Ethiopia stands first in area coverage and second in total annual production next to maize, and ranks the lowest yield compared with other cereals grown in Ethiopia [3]. At south nation nationality regional state (SNNPRS) the average yield was 1.49 t ha^{-1} and it is below the average yield of the country which was 1.75 t ha^{-1} [3]. This low yield and productivity is mainly due to the traditional farming system which is not supported by improved technologies such as proper sowing method and optimum seed rate[6] .The small size of teff seed postures problems during sowing and indirectly also in the weeding and threshing operations. At sowing, the very small seed size makes it difficult to control population density and even distribution. Farmers usually use higher seeding rates than those given by research recommendations, which may be due to their unclean seed with lower germination rate and apparently also to minimize weed infestation[7]. Hence, present teff production system is unable to satisfy the consumer's demand, since most Ethiopian farmers practice traditional farming system. Production system is not efficiently supported by modern technology due to research gap in choosing most feasible modern technology. Thus, maintaining seeding rate and sowing methods are one of the main challenges in teff production.

Seed rate is one of the important factors in achieving optimum level of plant density and has considerable effects on growth and development of crops [8 and 9]. In the study area, there is variation in using improved cultural practices like seeding rate and method of sowing and results in low yield teff. The most common method of sowing is broadcasting which is traditional and greatly reduces the amount of teff grain yield due to higher competition for resources and also lodging [10]. On the other hand, row planting maintains uniform population per unit area and provides easy access for carrying out cultural practices although requires more time, energy, and cost[11]. When the plant density exceeds an optimum level, competition among plants for light above ground and nutrients below ground becomes severe [5]. There was significant increase in yield components and yield of teff with decreasing seed rate from highest to lowest. On the other hand, the lodging percentage of the crop was increased by increasing the seed rate [12].

Most farmers practice the traditional sowing method by broad casting of the small seed at the rate of 25-55 kg /ha, which creates excess crop density and increases competition among plants for resources [6]. More over broadcasting methods requires additional seed rate compared to row sowing method thus increases cost of production. Furthermore, this sowing method results in lodging; which is the main cause for low yield of teff due to high plant density [12]. Row planting in teff is reported to have better yielding advantage over broadcast planting. To minimize the problem of lodging on teff, low seed rate, row planting, late sowing and application of plant growth regulators were used [13]. However, in the study area, there was lack of information on the response of teff crop to seeding rate and methods of sowing. Thus, increasing the productivity of teff through strategic manipulations of seeding rate and methods of sowing are paramount importance. Therefore, the research was initiated with the overall objective of evaluating the growth and yield response of teff to seeding rate and methods of sowing in southern Ethiopia.

2. MATERIALS AND METHODS

2.1 Description of the Study Area

The field experiment was conducted during 2018/19 main cropping season at Wolaita Sodo Agricultural Technical Vocational Education and Training (ATVET) College demonstration field in southern Ethiopia. An approximate geographical coordinates of the area is 6°34' N latitude and 37°43' E longitude having an altitude of 1883 m.a.s.l. The area is characterized with a bimodal rainfall distribution pattern with the total rainfall in the cropping season was 481.6 mm, and the mean maximum and minimum temperatures were 25.91°C and 15.35°C, respectively.

2.2 Treatments and Experimental Design

The two factors tested were seeding rate with five levels (2.5, 5, 10, 15, and 20kg ha^{-1}) and two methods of sowing (broadcasting and row). The treatments combined in factorial arrangement and laid out in randomized complete block design (RCBD) with three replications. The seeds were either drilled in specified rows or broadcasted by spreading on the plots. In the row sowing method, each plot contained 12 rows spaced at 20cm. All plots were 2m×2.4 m ($4.8m^2$) with 0.5m distance between plots in and 1.5m between blocks. Teff variety Tsedey (DZ-Cr-37) was used as a test crop. The variety was developed and released by Debrzeit Agricultural Research Center in 1984. It is a high yielding white-seeded with medium height and early mature cultivar with days to maturity of 82-90 [14]. The variety adapts to an altitude of 1500-2200 m.a.s.l and 150-200 mm annual rain fall during growing period with suitable temperature range 10-27 °C [15].

2.3 Agronomic Practices

The experimental field was prepared by using oxen plow and plowed four times before sowing. After plowing experimental field was pulverized and leveled to get smooth seed bed. Important agronomic practices applied as per their recommendation [2 and 16]. Nitrogen (N) at 41 kg/ha and phosphorous (P) at 46 kg P_2O_5 ha^{-1} in the form of urea (46-0-0) and di-ammonium phosphate (18-46-0) were applied. P was applied at sowing whereas N was applied into two split (at sowing and tiller stage). Planting was done in the main growing season following farmers planting time in the area (June 20, 2018/19). The experimental field was hand-weeded as required during the growing season.

3.DATA COLLECTED

3.1 Crop phenology

Days to 50% heading: Days to heading were recorded by counting the number of days from sowing to 50% of plants per plot exhibit panicles.

Days to 90% physiological maturity: It was recorded by counting the number of days from sowing when 90% of plants in a plot changed to light yellow color.

Lodging: The degree of lodging was measured just before the time of harvesting by visual observation based on the scales of 1-5 where scale no. 1 (0-15°) indicates no lodging, scale no. 2 (15-30°) indicate 25% lodging, scale no. 3 (30-45°) indicate 50% lodging, scale no. 4 (45-60°) indicate 75% lodging and scale no. 5 (60-90°) indicate 100% lodging[17]. The scales were determined by the angle of inclination of the main stem from the vertical line to the base of the stem by visual observation. Each plot was divided based on the displacement of the aerial stem in to all scales by visual observation. Each scale was multiplied by the corresponding percent given for each scale and average of the scales represents the lodging percentage of that plot.

3.2 Growth parameters

Plant height: It was measured at physiological maturity from the ground level to the tip of panicle from ten randomly selected plants in each plot.

Number of tillers per plant: The number of tillers per plant was recorded from the ten sample plant from a plot.

Effective tillers per plant: It was determined at physiological maturity from ten randomly selected plants per plot by counting the effective tillers per plant.

Yield components and yield

Panicle length: It is the length of the panicle from the node where the first panicle branches emerge to the tip of the panicle which was determined from an average of ten selected plants per plot.

Panicle number per plant: It was determined by counting for ten randomly selected plants per plot at physiological maturity.

Panicle seed weight: The average weights of the panicle at harvest, from ten randomly selected and pre-tagged plants were taken.

Thousand seed weight: It was counted after threshing at random from each plot and their weights were measured with sensitive balance and the weight was adjusted the grain moisture content to 12.5% by using grain moisture tester.

Grain yield: It was measured by taking the weight of the grains threshed from all six central rows of each plot and converted to kilograms per hectare after adjusting grain moisture content to 12.5%.

Statistical Data Analysis

All collected data were subjected to analysis of variance (ANOVA) using Gestate software statistical computer software [18]. Least Significance Difference (LSD) test at 5% level of probability was performed when significance differences among treatment means were detected.

4. RESULTS AND DISCUSSION

4.1 Days to panicle emergence

Days to panicle emergence of teff was affected significantly ($p \leq 0.05$) by the main effects of both sowing methods and the interaction effects of sowing methods and seed rates (Table 1). The longest (57.27) days to 50% panicle emergence were recorded at row planting with seed rate of 2.5 kg ha^{-1}. The shortest (41.47) days to 50% panicle emergence were recorded from broadcasting method under higher seed rate (20kg ha^{-1}). The earlier days to 50% panicle emergence of plants under higher seed rate at broadcasting method might be attributed to the presences of competition for growth resources that made them to shift their phenology from vegetative to reproductive stage while plants grown at row planting with lower seed rate become free of competition for resources and get longer time to reach reproductive stage. The delayed days to panicle emergence with at a lower seed rate might be related with reduced competition among plants while at higher seed rate there was a strong competition for capturing of growth resources which in turn brings stress on the plant and make them shift their vegetative growth stage to reproductive[19].

Table 1: Interaction effects of sowing method and seed rate on days to panicle emergence of teff

Treatment		
Sowing method	Seed rate(Kg ha^{-1})	Panicle emergence
Row	2.5	57.27^{e}
	5	55.86^{de}
	10	54.14^{de}
	15	52.26^{cd}
	20	45.36^{ab}
Broadcast	2.5	48.67^{bc}
	5	46.45^{b}
	10	45.63^{ab}
	15	44.78^{ab}
	20	41.47^{a}
	LSD (0.05)	
Sowing method(Means over all seed rates)		
	Row	53.97^{a}
	Broadcast	45.4^{b}
	LSD (0.05)	2.651
	CV (%)	0.5

Means followed by the same letter within a column within the same treatment category are not significantly different at 5% level of significance. CV = Coefficient of variation; LSD = Least significant difference

4.2. Days to Physiological Maturity

Differences in the time of crop maturity are caused by the genetic makeup of the variety or by environmental conditions existing during their growth, grain filling or harvesting period of the crop [20]. Row sowing at 2.5 kg ha^{-1} seed rate prolonged the crop significantly longer time to 90% physiological maturity over other treatments. Teff plants grown under higher seed rate for both sowing methods took shorter time to reach at their physiological maturity as compared to lower seed rate (Table 2). Similar finding was reported by in which treatments under lower seed rate sown in rows took longer time to reach at their physiological maturity [21]. This might be due to less intra-specific competition of plants resulted from reduced seed rate and better management of plants in the rows that contributed to fair utilization of growth resources in the soil.

4.3. Plant Height

Plant height was significantly affected by the main effects of sowing methods and seed rates ($P \leq 0.01$) whereas their interaction effects was not significant (Table 2). Plant height was found to be more due to row sowing over broadcast sowing method. Similarly, there was increase in plant height from highest to the lowest seed rate .Tallest plant height (87.67cm) was obtained at seeding rate of 2.5kg ha^{-1} whereas the minimum plant height (81.83cm) was recorded from the higher seeding rate of 20kg ha^{-1} (Table 2). This might be due to high competition between plants for growth resources which led to less vegetative growth and plant height. Similar to this finding reported increased plant height due to row sowing method and lower seed rate [21]. Similarly reported that plant height increased as seeding rate decreased [19]. Taller and more branched plants were observed at the lower plant densities of sesame [22]. Seeds planted using row method gave 1.5% taller plant height (84.93 cm) than sown using broadcasting. This might be due to less intraspecific competition of plants for light and nutrients as well as soil moisture which allows for increased vegetative growth. Related findings were reported by other researchers who showed that an increase in plant height with row method of sowing and lower seeding rate [21].

4.4. Panicle Length

Panicle length was significantly ($p \leq 0.01$) affected by sowing method and seeding rate but their interaction effect was non-significant (Table 2).Panicle length is one of the major yield attributes of teff that is positively associated with grain yield[23]. Panicle length decreased as the seeding rate increased from 2.5 to 25kg ha^{-1}(Table 2). The increased panicle length from the combination of row sowing and reduced seeding rate might be the result of more space provided for the crop to utilize more growth resources by decreasing competition among plants. This finding is similar with who reported that significantly higher panicle length was observed under low seeding rate than in high seeding rate [24]. Seeds planted using the row method of sowing produced longer panicle length (34.45cm), while broadcasted seeds gave shorter panicle length (31.38cm) (Table 2).This could be due to less competition in row sown teff. Similarly, the authors of indicated increment in panicle length with row method of sowing as compared to broadcast sowing method [25].

4.5. Panicle number

Numbers of panicles per plant were increased significantly as seed rate decreased from 20 to 2.5 kg ha^{-1} (Table 2). 2.5 kg ha^{-1} seed rate produced the highest number (18.71) while 20 kg ha^{-1} seed rate produced the least panicle number (12.90). Also increased number of panicle recorded at row planting (17.42) while lowest number of panicle obtained from broadcasting method (13.94) (Table 2). Increased number of panicle at the lower seeding rate was attributed to more interception of sunlight for photosynthesis and resulted in production of more assimilate for partitioning towards the development of more panicle number

The higher number of panicle per plant might be from the effect of more space and nutrients utilization by the plants. In agreement with this result reported that panicle number increased with decreasing seed rate [26].

4.6. Thousand Seed weight

The analysis of variance of the data showed significant ($p \leq 0.05$) variation in thousand seed weight for the main effect of seed rates and sowing methods but not for the interaction effects. In general, as seeding rate decreased, thousand-seed weight increased and this might be because lower seeding rate enhances efficiently utilization of the existing resources and thus improves vegetative as well as reproductive growth. The heaviest thousand seed weight (0.38g) was obtained at the seed rate of 2.5 kg ha^{-1}whereas the lightest thousand seed weights (0.32g) were obtained in response to founding at the highest seed rates of 20 kg ha^{-1}(Table 2). Like finding was reported by who revealed a higher thousand-seed weight with decreasing seeding rate which is due to vigorous crop growth compared to reduced thousand-seed weight and final yield at a higher seeding rate [27]. Thousand-seed weight was also influenced by the methods of sowing where plants sown in rows produced higher thousand-seed weight than the broadcast method. Furthermore, seeds planted in row method of sowing showed 12.5% increase in thousand-seed weight over broadcasting (Table 2). This might be due to increased seed content and weight as a result of efficient resource utilization.

4.7. Panicle seed weight

Panicle seed weight was significantly affected by the main effects of sowing methods and seed rates ($P \leq 0.01$) whereas their interaction effects was not significant (Table 2). Decreasing the seeding rate significantly increased panicle seed weight. The maximum of panicle seed weight (1.43g) was recorded at the lowest seed rate of 2.5 kg ha^{-1} whereas minimum panicle seed weight (0.74g) was recorded at the seed rates of 20 kg ha^{-1}. The greater panicle seed weight was obtained from row planting than broadcasting method. This result indicated that panicle seed weight decreased with the increasing seeding rate and determined the superiority of lower seeding rate against the larger seeding rates in enhancing of panicle seed weight.

When seed rate increased, planting density also increased significantly, leading to inflexible competition among plants for growth factors. Supporting the results of this study reported that low seeding rate in rice generally increased panicle seed weight; however, fewer and smaller kernels per spike can occur which results in little change in total grain yield [28].

Table 2: Days to 90% physiological maturity, plant height, panicle length, panicle number, panicle seed weight and thousand seed weight of teff as affected by sowing methods and seed rates of teff grown

Sowing method	PM	PH(cm)	PL(cm)	PN(no)	PSW(g)	TSW(g)
Row sowing	84.93^{b}	90.57^{b}	34.45	17.42^{b}	1.400	0.36^{b}
Broadcast sowing	83.67^{a}	81.42^{a}	31.38	13.94^{a}	0.810	0.32^{a}
LSD(0.05)	0.785	2.188	0.799	0.526	0.0878	0.02116
Seed Rates (kg ha^{-1})						
2.5	87.67d	90.50^{c}	35.90^{d}	18.71^{d}	1.430^{d}	0.381^{c}
5	85.00c	88.00^{bc}	33.86^{c}	16.62^{c}	1.272^{c}	0.3633^{b}
10	84.17^{bc}	85.16^{b}	32.65^{bc}	15.35^{b}	1.088^{b}	0.3433^{ab}
15	82.83^{ab}	85.55^{b}	31.89^{ab}	14.83^{b}	0.992^{b}	0.3233^{a}
20	81.83^{a}	80.77^{a}	30.30^{a}	12.90^{a}	0.743^{a}	0.3100^{a}
LSD (0.05)	1.242	3.459	1.263	0.831	0.1388	0.03346
CV (%)	1.2	3.3	3.2	4.4	10.4	8.1

Means followed by the same letter within a column within the same treatment category are not significantly different at 5% level of significance. CV = Coefficient of variation; LSD = Least significant difference

4.8. Lodging percentage

Lodging is the state of permanent displacement of the stems from their upright position [8]. It can be induced by both external and internal factors like wind, rain and morphological traits of the crops or by their interactions. Described the types of lodging as stem lodging and root lodging [29]. The grain bearing organs of cereals are found at the top of the stems and therefore, exert a strain on the stalk especially under high wind or rain. Lodging index was affected highly significantly ($P \leq 0.01$) by the main effects of both seed rates and sowing methods, and significantly ($P \leq 0.05$) by the interaction effects of sowing methods and seed rates (Table 3). The higher Lodging percentage was found under the maximum seed rate and both sowing methods. The higher lodging at higher seed rates under both planting methods might be due to the fact that the highest seed rate at both broadcast and row sowing method enhanced fast vegetative growth and succulent stem elongation of teff. On the other hand, lower lodging index (30.19%) at lower seed rate (2.5kg ha^{-1}) under row planting might be related to the vigorous and strong stem of teff plant due to less density stress. In general with this result and reported that low plant density teff could reduce lodging [30 and 10]. Also showed decreasing of lodging in teff crop was observed from decreasing the seed rate and planting in row [21]. The tendency of plant lodging at increased seed rate is also associated with a disorganized light profile which leads to increased weak and thin plant height which is highly susceptible to various climatic situations [31]. The incidence of lodging is also found to be reduced, as the stem of teff is better able to support the weight of the filled head of grain [6].

4.9. Number of Total and Effective Tillers per Plant

The number of total and particularly that of effective tillers per plant is the most important yield component because the final yield is mainly a function of panicle-bearing effective tillers per unit area. As the number of total and effective tillers per plant increases, the grain yield of crops also increases. As the seed rate increases, the numbers of total and effective tillers decreases and vice versa [32].

Seed rate and sowing method interaction significantly ($p \leq 0.01$) affected on the number of both total and effective tillers per plant (Table 3). The highest total and effective tiller number per plant was observed from row planting with seed rate of 2.5 kg ha^{-1} while the lowest total tiller number per plant was recorded at broadcasting method with seed rate of 20 kg ha^{-1}. The maximum total and effective tiller number per plant at lower seed rate and row planting might be attributed to better utilization of growth resources [33].

4.10. Grain yield

Grain yield was significantly ($P \leq 0.01$) differed due to the main and interaction effects of seed rate and sowing methods (Table 2). Crop yield is a function of a number of factors and processes such as amount of light intercepted by the canopy, metabolic efficiency of plants and the translocation efficiency of photosynthates from leaves to economic parts. These processes are affected by plant densities. The highest grain yield (2.5 t ha^{-1}) was obtained from the lower seeding rate of 2.5 kg ha^{-1} combined with row sowing method. The greater yield obtained from lower seeding rate combined with row planting might be due to the combined effects of row sowing method that facilitated better field management and lower seeding rate that contributed to lesser plant population by minimizing intraspecific competition for growth resources among plants. This result is similar to who reported the combination of row sowing method and lower seeding rate gave the highest grain yield of teff [25]. Similarly, also revealed that there was significant increase in yield and yield components of teff with decreased seeding rate from the highest to the lowest [12]. Likewise, stated that tef sown at lowest seed rate (10 kg ha^{-1}) gave the highest grain yield as compared to the yield from the highest seed rates [4]. Maximum yield for any plant in a crop may be achieved at that density of plants at which competition with the plant is minimal [35].

5.CONCLUSION AND RECOMMENDATION

Days to physiological maturity, plant height, panicle length, panicle number, panicle seed weight and thousand-seed weight were not affected by the interaction effect of seeding rate and methods of sowing. But, the interaction of the two factors significantly affected days to panicle emergence, total tillers, and effect tillers, lodging percentage and grain yield. In general, the combination of row sowing method with 2.5kg ha^{-1} seed rate was the best treatment for high grain yield of teff at the study site. However, it is advisable to undertake further research across soil type, years and locations to draw sound recommendation on a wider scale.

Table 3: Interaction effects of sowing method and seed rate on lodging percentage, number of total tillers per plant, effective tillers per plant and grain yield of teff grown

Treatment					
Sowing method	Seed rate(Kg ha^{-1})	Lodging%	No. of total tillers $plant^{-1}$	No. of effective tillers $plant^{-1}$	Grain yield(t ha^{-1})
Row	2.5	30.19^{a}	21.60^{h}	17.373^{f}	2.50^{g}
	5	39.26^{b}	17.61^{g}	12.997^{e}	1.96^{f}
	10	41.95^{bc}	15.42^{f}	9.547^{d}	1.63^{e}
	15	44.70^{cd}	12.04^{e}	6.830^{c}	1.38^{d}
	20	48.76^{de}	10.66^{d}	5.117^{b}	1.17^{c}
Broadcast	2.5	52.44^{ef}	11.17^{de}	6.723^{c}	1.61^{e}
	5	53.22^{fg}	8.33^{c}	5.320^{b}	1.38^{d}
	10	54.64^{fg}	6.43^{b}	4.083^{a}	1.20^{c}
	15	57.51^{g}	4.74^{a}	3.807^{a}	0.99^{b}
	20	63.18^{h}	3.87^{a}	3.150^{a}	0.76^{a}
	LSD (0.05)	4.338	0.965	0.960	1.428
Sowing method(Means over all seed rates)					
	Row	40.97^{a}	15.47^{b}	10.37^{b}	1.73^{a}
	Broadcast	56.20^{b}	6.91^{a}	4.62^{a}	1.19^{b}
	LSD (0.05)	1.940	0.431	0.429	0.638
	CV (%)	5.2	5.0	7.5	5.7

Means followed by the same letter within a column within the same treatment category are not significantly different at 5% level of significance. CV = Coefficient of variation; LSD = Least significant difference

ACKNOWLEDGMENTS

The authors would also like to express their appreciation to Wolaita Sodo Agricultural TVET College for providing them the necessary materials and experimental site for carrying out the research.

REFERENCES

1. Berhe, T. and Zena, N., 2008. Results in a trial of system of tef intensification at Debre Zeit, Ethiopia.

2. Tesfay, T. and Gebresamuel, G., 2016. Agronomic and economic evaluations of compound fertilizer applications under different planting methods and seed rates of tef [*Eragrostis tef* (Zucc.)Trotter] in northern Ethiopia. *Journal of Dry lands*, 6(1): 409-422.

3. CSA (Central Statistic Authority), Agricultural Sample Survey: Report on Area and Production of Major Crops (Private Peasant Holdings"meher"Season), Vol. I, CSA, Addis Abeba, Ethiopia, 2017/18.

4. Tefera, H. and Ketema S., 2001. Production and importance of tef in Ethiopian Agriculture. Pp. 3-7. In: Hailu Tefera, Getachew Belay and Mark Sorrels (eds). Narrowing the rift. Tef Research and Development Proceeding of the International Workshop on Tef genetics and Improvement, Debrzeit, Ethiopia, 16-19 October 2000.

5. Baloch, A.W., A.M. Soomro, M.A. Javed, M. Ahmed, H.R. Bughio, M.S. Bughio and N.N. Mastoi. , 2002. Optimum plant density for high yield in rice (*Oryza sativa* L.). *Asian Jour nal.Plant Science.* 1: 25-27.

6. Berhe, T., 2010. Productivity of teff [*Eragrostis tef* (Zucc.) Trotter]. New approach with dramatic results (Unpublished). Report Addis Ababa, Ethiopia.

7. Ketema, S., 1997. Teff [*Eragrostis tef* (Zucc.) Trotter]. Promoting the conservation and use of underutilized and neglected crops. Biodiversity institute, Addis Ababa, Ethiopia.

8. Ketema, S., 1993. Phenotypic variations in teff (*Eragrostis tef*) germplasm- morphological and agronomic traits. A catalon Technical manual No. 6. Institute of Agricultural Research, Addis Ababa, Ethiopia.

9 .Kurubetta, K. D., 2006. Effect of time of sowing, spacing and seed rate on seed production potentiality and quality of fodder cowpea (*Vigna unguiculata* (L.) Walp]," M.Sc. Thesis, College Of Agriculture, Dharwad University of Agricultural Sciences, Dharwad, India.

10. Hundera F, Bogale T, Tefera H, Assefa K, Kefyalew T. ,2001. Agronomy research in tef. In: Hailu Tefera, Getachew Belay and Sorrells M (eds) Narrowing the Rift. Teff Research and Development. Proceedings of the International Workshop on Tef Genetics and Improvement, Debre Zeit, Ethiopia, pp: 167-176.

11. Chandrasekaran, B., Annadurai, K., and Somasundaram, E., 2010. *A textbook of agronomy*. New Age International Limited.

12. Mitiku, M., 2008 .Effects of seeding and nitrogen rates on yield and yield components of tef [Eragrostis tef (Zucc.) Trotter]. Haramaya University, Haramaya, Ethiopia.

13. Tesfahun, W., 2018. "Teff Yield Response to NPS Fertilizer and Methods of Sowing in East Shewa, Ethiopia." Journal of Agricultural Sciences–Sri Lanka 13, no. 2 (2018): 162-173.

14. Tiruneh, K., 2000. Genotype × Environment Interaction in Tef. In: Hailu, T., Getachew, B. and Sorrells, M. (eds.). Narrowing Rift: Tef Research and Development. Proceeding of International Workshop on Tef Genetics and Improvement, October 2000, 45pp, Addis Ababa, Ethiopia.

15. (MoARD) Ministry of Agriculture and Rural Development. , 2007." Crop Development Department. Crop Variety Register". Issue No. 10. Addis Ababa, Ethiopia.

16. Tesfahunegn, G.B., 2015. Short-term effects of tillage practices on soil properties under Tef [*Eragrostis tef* (*Zucc. Trotter*)] crop in northern Ethiopia. *Agricultural Water Management*, 148: 241-249.

17. Donald, L.S., 2004. Understanding and reducing lodging in cereals. Advances in Agronomy 84:217–271.

18. VSN International Ltd., 2012. Genstat Procedure Library Release. 15th Edition. https://www.vsni.co.uk/.

19. Sahle S, Altaye T., 2016. Effects of Sowing Methods and Seed Rates on Yield Components and Yield of Teff in Soro Woreda, Hadya Zone, Southern Ethiopia. Journal of Natural Sciences Research.6(19):2225-0921.https://iiste.org/Journals/index.php/JNSR/article/view/33774/34718

20. Geng, S., 1984. Effect of Variety and Environment on head Rice yield. 5th annual report to the California rice growers, food and agricultural state of California, USA.

21. Refissa, L., 2012. Effects of sowing method and inorganic fertilizer type on yield and yield components of teff [*Eragrostis tef* (zucc.) Trotter] at Guduru Woreda, Western Oromia, Ethiopia. M.SC. Thesis, Haramaya University, Haramaya, Ethiopia.

22. Caliskan, S., Arslan, M., Arioglu, H.,and Isler, N., 2004. Effect of planting method and plant population on growth and yield of sesame (*Sesamum indicum* L.) in a Mediterranean type of environment. *Asian Journal of Plant Science*, 3(5), 610-613.

23. Tefera, H., Ketema, S. and Tesemma, T.E.S.F.A.Y.E., 1990. Variability, heritability and genetic advance in tef (Eragrostis tef (Zucc.) Trotter) cultivars. *Variability, heritability and genetic advance in tef (Eragrostis tef (Zucc.) Trotter) cultivars. 67*(4), pp.317-320.

24. Berhe, T., 2008. Increasing Productivity of Teff [*Eragrostis tef* (Zucc.)] Trotter: New Approaches with Dramatic Results (Unpublished Report)" Addis Ababa, Ethiopia.

25. Shiferaw, T., 2012. Effects of sowing method and inorganic fertilizer type on yield and yield components of teff [*Eragrostictef* (Zucc.) Trotter] at A'ada Woreda, Central Ethiopia, M.Sc. Thesis, Haramaya University, Haramaya, Ethiopia.

26. Sahle, S. and Altaye T., 2016. Effects of Sowing Methods and Seed Rates on Yield Components and Yield of Tef in Soro Woreda, Hadya Zone, Southern Ethiopia. *Journal of Natural Sciences Research,* 6 (19):2225-0921.

27. Mazurek, J. and Sabat, A., 1984. Effect of sowing rate and nitrogen fertilizer application on yields of some triticale cultivars. Pamie hook tnik Pulawski, 83, pp.85-93.

28. Reda, A., Dechassa, N. and Assefa, K., 2018. Evaluation of seed rates and sowing methods on growth, yield and yield attributes of tef [Eragrostis tef (Zucc.) Trotter] in Ada District, East Shewa, Ethiopia. *American-Eurasian J. Agric. & Environ. Sci, 18*(1), pp.34-49.

29. Berry, P.M., Sterling, M., Spink, J.H., Baker, C.J., Sylvester-Bradley, R., Mooney, S.J., Tams, A.R. and Ennos, A.R., 2004. Understanding and reducing lodging in cereals. *Advances in agronomy*, *84*(04), pp.215-269.

30. Assefa, K., 1991. *Effects of some synthetic plant growth regulators on lodging and other agronomic and morphological characters of tef [Eragrostis tef, (Zucc.) Trotter]* (Doctoral dissertation, M. Sc. Thesis, Alemaya University of Agriculture, Alemaya, Ethiopia. DOI: http://dx. doi. Org/10.1111/j. 1439-0523.2010. 01782. X).

31. Khalil, S.K., Wahab, A., Rehman, A., Muhammad, F., Wahab, S., Khan, A.Z., Zubair, M., Shah, M.K., Khalil, I.H. and Amin, R., 2010. Density and planting date influence

phenological development assimilate partitioning and dry matter production of faba bean. *Pak. J. Bot, 42*(6), pp.3831-3838.

32. Soomro, U.A., Rahman, M.U., Odhano, E.A., Gul, S. and Tareen, A.Q., 2009. Effects of sowing method and seed rate on growth and yield of wheat (*Triticum aestivum*). *World Journal of Agricultural Sciences, 5*(2), pp.159-162.

33. Abebe, L., 2013. Effects of planting method, fertilizer type and seed rate on growth and productivity of teff [(*Eragrostis tef* (Zucc.) Trotter)].M.Sc. Thesis, Hawassa University, Ethiopia.54 p.

34. Laekemariam, F., Gidago, G. and Taye, W., 2016.'Lowering tef seed rate using a seed spreader via the participatory approach in south Ethiopia'. Advances in Life Science and Technology, 5:2224-7181

35. Helenius, J. and Jokinen, K., 1994. Yield advantage and competition in intercropped oats (*Avena sativa* L.) and faba bean (*Vicia faba* L.): Application of the hyperbolic yield-density model. *Field Crops Research, 37*(2), pp.85-94.